THEATER MISSILE DEFENSE IN JAPAN: IMPLICATIONS FOR THE U.S.-CHINA-JAPAN STRATEGIC RELATIONSHIP

Patrick M. O'Donogue

September 2000

*****

*****

Comments pertaining to this report are invited and should be forwarded to: Director, Strategic Studies Institute, U.S. Army War College, 122 Forbes Ave., Carlisle, PA 17013-5244. Copies of this report may be obtained from the Publications and Production Office by calling commercial (717) 245-4133, FAX (717) 245-3820, or via the Internet at rummelr@awc.carlisle.army.mil

****

Most 1993, 1994, and all later Strategic Studies Institute (SSI) monographs are available on the SSI Homepage for electronic dissemination. SSI's Homepage address is: http://carlisle-www.army.mil/usassi/welcome.htm

*****

The Strategic Studies Institute publishes a monthly e-mail newsletter to update the national security community on the research of our analysts, recent and forthcoming publications, and upcoming conferences sponsored by the Institute. Each newsletter also provides a strategic commentary by one of our research analysts. If you are interested in receiving this newsletter, please let us know by e-mail at outreach@awc.carlisle.army.mil or by calling (717) 245-3133.

ISBN 1-58487-035-4

FOREWORD

This study, written by Colonel Patrick M. O'Donogue, USMC, as a strategy research project, considers a topic of key importance to U.S. national security—Japan's agreement to cooperate with the United States on the development of a theater missile defense (TMD). China vigorously opposes this plan and insists that U.S. development of TMD and national missile defense systems will destabilize its strategic relationship with the United States. China's concerns center primarily on Japan's and Korea's development of deployable upper-tier missile defense capabilities and technology, along with equipment transfers of any kind to Taiwan.

Colonel O'Donogue, a member of the Army War College Class of 2000, does a remarkable job of addressing this matter in a thoughtful, articulate, and comprehensive manner. The Strategic Studies Institute is pleased to publish his study as part of our Letort Paper series.

DOUGLAS C. LOVELACE, JR.
Director
Strategic Studies Institute

BIOGRAPHICAL SKETCH OF THE AUTHOR

PATRICK M. O'DONOGUE, a colonel, U.S. Marine Corps, is the Assistant Chief of Staff, G-7, I Marine Expeditionary Force. From August 1999 to June 2000 he was a student at the U.S. Army War College. His military career began when he was commissioned a second lieutenant upon graduation from Oregon State University. Lieutenant Colonel O'Donogue is a Naval aviator and served in Operations DESERT SHIELD, SOUTHERN WATCH, and SILENT ASSURANCE. He earned an M.A. in National Strategy and Strategic Studies from the Naval War College in 1994.

THEATER MISSILE DEFENSE IN JAPAN: IMPLICATIONS FOR THE U.S-CHINA-JAPAN STRATEGIC RELATIONSHIP

Introduction.

The role of theater missile defense (TMD) in the U.S.-China-Japan strategic relationship has taken on increased importance as its development and deployment by the United States and, most likely, Japan becomes inevitable. The August 1999 agreement between the United States and Japan to cooperate on research and development of a sea-based TMD system raised the stakes in what has been, so far, a "war of words" on the need for TMD in Japan, between the United States and Japan, on one side, and China, on the other. American and Chinese perceptions of the efficacy and utility of theater missile defense in national security planning differ radically. U.S. officials and scholars argue that missile defense promotes national and international security. The Chinese, however, reject the assumptions and logic of U.S. support for national missile defense (NMD) and TMD.

U.S. nuclear deterrent strategy relies on both the U.S. strategic nuclear triad and U.S. ballistic missile defenses, with the latter reinforcing the stabilizing effects of deterrence. Unlike the United States, China relies exclusively on land-based missiles as deterrents. So China perceives a strengthened U.S. defense against her land-based missiles as destabilizing, because such defenses deprive China of her power to deter. However, U.S. officials perceive a growing missile threat from "rogue nations," while the threats of accidental and unauthorized launches remain. The United States also has an inherent obligation to protect forward-based and forward-deployed forces against the threat of ballistic missile attacks. The 1998

North Korean Taepo Dong-1 launch gave credibility to such threats.[1]

Chinese officials argue, however, that these threats are not serious enough to justify NMD and TMD development.[2] They insist NMD and TMD pose serious negative implications for global arms control and nonproliferation efforts, specifically the Strategic Arms Limitation Talks (SALT) II, the Comprehensive Test Ban Treaty (CTBT), the Missile Technology Control Regime (MTCR), and the Fissile Material Cut-off Treaty (FMCT). U.S. officials counter that arms control and nonproliferation are in the interests of the community of nations, and are not threatened by NMD and TMD.[3]

Growing technological cooperation between the two nations means inevitable Japanese help with development and eventual deployment of TMD. At a minimum, lower-tier TMD deployments in Japan will be continually upgraded to protect U.S. troops and military assets. The Chinese have expressed a willingness to accept lower-tier TMD deployment that protects U.S. bases. But China opposes the development and deployment of upper-tier TMD systems, especially sea-based versions, which could be employed to protect Taiwan.[4] Shu Zukang, Director General of the Chinese Foreign Ministry's Department of Arms Control, best expresses the Chinese position:

> We do not envisage a dispute concerning development of what we would call genuine TMD. Here I am referring to those anti-theater missile systems used solely in a limited area. What China is opposed to is the development, deployment and proliferation of antimissile systems with potential strategic defense capabilities in the name of TMD that violate the letter and spirit of ABM and go beyond the legitimate self-defense of relevant countries.[5]

This statement lends *de facto* legitimacy to the sale of Patriot PAC-3 missiles, a lower-tier capability, to Taiwan. The Chinese insist, however, that Taiwan's status as part of China obviates its entitlement to the independent receipt of arms. But China seems confident that it maintains sufficient force, both in aircraft and ballistic missiles, to overwhelm any Taiwanese defense, which renders such sales more an issue of principle than practicality.

Japanese adoption of TMD introduces new dynamics into the U.S.-Japan alliance. If brought to fruition in a rational way, TMD will precipitate a broad range of operational initiatives, will require changes to bilateral doctrine and will redefine the role of the Japanese Self-Defense Forces (JSDF). TMD poses the same implications for bilateral integration at the operational and support levels as the Defense Guidelines do, with even more tangible consequences for the bilateral alliance. TMD will undoubtedly influence the way the United States and Japan plan, procure, consult, and operate in the future.[6]

This study describes the historical precedence that led to a decision to develop TMD and seriously consider its deployment in Japan. It then explores the strategic setting that shapes the positions of Japan, China and the United States with regard to TMD. Next, it analyzes possible effects of TMD on the U.S.-Japan alliance. It then reviews Chinese objections to TMD and their implications for the U.S.-China relationship. Following this, it explores the U.S. options for TMD development and deployment. The study concludes with a discussion of the desirability of the United States maintaining a firm, open dialogue with China that addresses cogent concerns that, if carefully handled, offer reasons for optimism.

Background.

Theater missile defense encompasses a range of lower-tier (endo-atmospheric–within the atmosphere) and upper-tier (exo-atmospheric–outside the atmosphere) systems intended to intercept and destroy short (<1000 km) and medium range (1000-3500 km) ballistic missiles. (See Table 1.) Relevant lower-tier U.S. systems under development include the U.S. Navy Aegis Standard Missile

II, also known as Navy Area Wide, and the U.S. Army Patriot missile, PAC-3 variant (Table 2). These systems advertise effectiveness against missiles with up to a 1,500-km range.

They do not fall under the provisions of the 1972 Anti-Ballistic Missile (ABM) treaty.[7] Upper-tier TMD

Type Missile	Range
Short Range Ballistic Missile (SRBM)	<1,000 km
Medium Range Ballistic Missile (MRBM)	1,000 km to 3,500 km
Intermediate Range Ballistic Missile (IRBM)	3,500 km to 5,500 km

Table 1. Types of Ballistic Missiles.

System	Type	Missile
Patriot PAC-3	Lower-tier	Patriot
Navy Area Wide	Lower-tier	Standard Missile II Block IVA
Medium Extended Air Defense System (MEADS)	Lower-tier	MEADS
Theater High Altitude Air Defense (THAAD)	Upper-tier	THAAD

Table 2. U. S. Ballistic Missile Defense Systems.[8]

systems are in various stages of development. They include the Navy Theater Wide system, again, an Aegis-based system using the Standard Missile III and the U.S. Army Theater High Altitude Air Defense (THAAD) system. (See Table 2.) Both systems would incorporate hit-to-kill technology in an exo-atmospheric vehicle to intercept longer range, higher speed missiles.[9]

U.S.-Japan dialogue on missile defense, especially theater missile defense, is well-established. Limited discussions began as early as 1983, at the inception of the Strategic Defense Initiative ("Star Wars").[10]Given the impetus of the 1990-91 Gulf War and U.S. experience with Iraqi SCUD ballistic missiles, in June 1992 the Bush Administration proposed Japanese (and South Korean) TMD deployments. Secretary of Defense Les Aspin visited Tokyo in October 1993 and asked for formal Japanese participation in TMD. But trade tensions between the two countries limited the U.S. request for participation to only three options: joint development; buying TMD "off-the-shelf" (primarily from U.S. manufacturers); or engaging in gradual technology exchanges.[11] This approach met with a cool reception in Japan. In fact, the Japanese perceived it as a blatant U.S. attempt to obtain their technology.

The TMD alliance revived in 1994 with the establishment of the bilateral U.S.-Japan Theater Missile Defense Working Group. The limited U.S. offers of 1993 faded from view while the partners concentrated on examining Japanese missile defense requirements and the potential for bilateral collaboration.[12] But several years of discussion produced no impetus to move talks beyond the exploratory phases. Then, in 1996, developments in North Korean nuclear and ballistic missile programs and the steady growth of Chinese military capabilities, specifically a significant ballistic missile force and its demonstrated use in the Straits of Taiwan, proved to be a catalyst: the United States and Japan began serious TMD negotiations. The event that most riveted Japanese attention to the threat from ballistic missiles, however, was the North Korean test of a three stage Taepo Dong-1 missile (See Table 3) in August 1998.

Missile System	Range/Payload
SCUD B	300 km/single conventional or chemical warhead
SCUD C	550 km/single conventional warhead
Nodong-1	1,300 km/single conventional chemical or nuclear warhead
Nodong-2	1,500 km/single conventional chemical or nuclear warhead
Taepo Dong-1	2,000 km/single conventional or nuclear warhead

Table 3. North Korean Ballistic Missiles. [13]

The Japanese concluded that the most logical program for bilateral technology cooperation would be a Navy Theater Wide (NTW) program. The JSDF's Maritime Self-Defense Force possessed the necessary platforms (Aegis cruisers), and the program lacked sufficient maturity that would guarantee meaningful participation by Japanese industry. This pragmatic approach to development maximized potential profits and assured favorable bilateral technology cooperation for Japanese industry. The U.S. impetus for pressing the Japanese on TMD was the presence of over 100,000 U.S. military personnel in Northeast Asia, with 55,000 of those stationed in Japan (to include Okinawa). The test of the North Korean Taepo Dong-1 ballistic missile sharpened Japanese public attention and underscored the apparent need for protection from unpredictable nations. Thus in August 1999 the Japanese joined the United States in a 5-year defense program to cooperate on research and development of a theater missile defense system.[14]

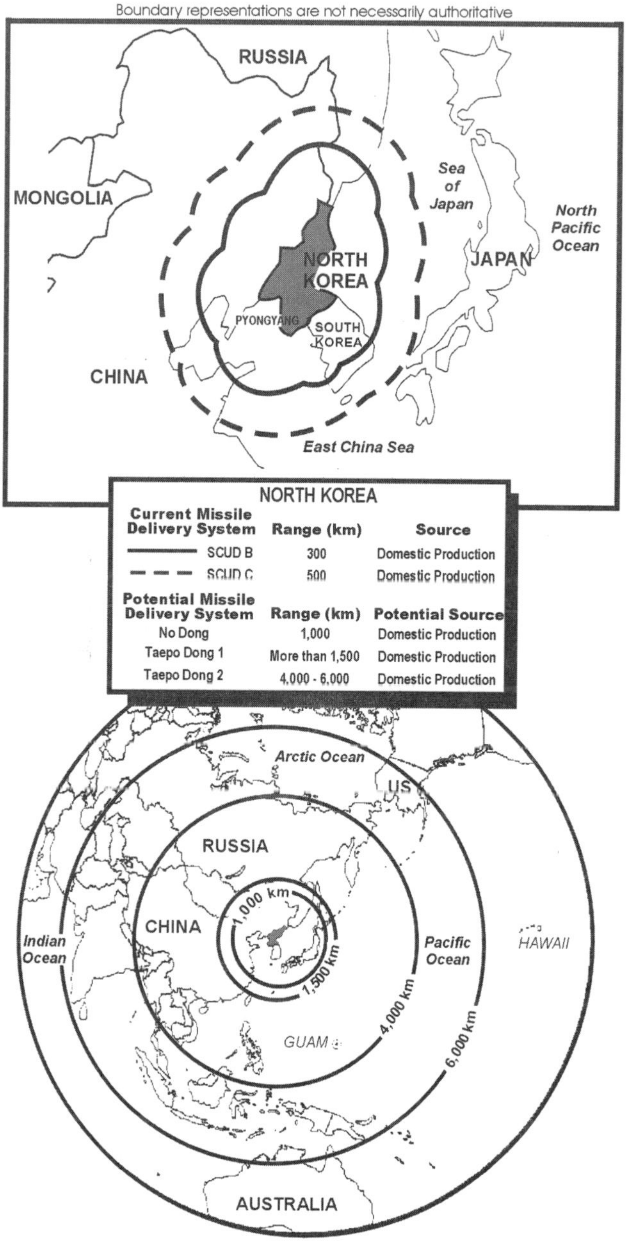

Figure 1. Estimated Ranges of Current and Potential North Korean Ballistic Missiles.[15]

Likewise, the Japanese became more aware of the Chinese ballistic missile program. The Chinese threat did not seem as immediate as the one from North Korea; however, Japanese officials realized that missiles directed at Taiwan could be redirected toward them and that road-mobile missiles now under development in China could be targeted against Japan. Should the Japan-China relationship change, the Japanese knew the risks.

The Strategic Setting.

East Asia in general, and Northeast Asia in particular, appears to be in a dangerous strategic setting. The regional characteristics include major shifts in the balance of power, skewed distributions of economic and political power within and between countries, political and cultural diversity, an anemic security apparatus, and widespread territorial disputes over natural resource issues.[16] Chances for escalation of tension in the area seem great, especially without a U.S. military presence in the region. The 20th century history of East Asia proves that mistrust between two or more potential adversaries can lead each side to take precautionary and defensively-motivated measures that are perceived as offensive threats.[17] These perceptions can lead to countermeasures in kind, begetting an action-reaction cycle that increases regional tensions and creates a self-fulfilling prophecy about the region's volatility.[18] In view of all these variables that might fuel insecurities, East Asia appears quite dangerous. Not only could dramatic and unpredictable changes in the distribution of military capabilities increase uncertainty and mistrust, but the importance of sea-lanes and secure energy supplies to almost all regional actors could encourage a destabilizing competition to develop power-projection capabilities on the seas and in the skies. Because of the perception as offensive threats, power-projection forces heighten spiraling tensions as opposed to weapons that can only defend a nation's homeland. Perhaps even more important in East Asia than these variables are psychological factors (such as the historical mistrust and animosity among regional actors) and political geography issues relating to the Taiwan question, which make defensive systems (such as TMD) in the region appear threatening to other nations' security.[19]

Policy Concerns.

Japanese policy seeks to negotiate the delicate balance among its own security; its political, diplomatic, and economic alliance with the United States; and its political, diplomatic, and economic relationship with China. Japan still depends on American strategic security, although some would argue this relationship approaches "normalcy."[20]So Japan's long-term policy goal may be self-sufficiency, even in matters of security. But would a TMD alliance between the United States and Japan hasten self-sufficiency or delay it? Other questions quickly arise: Is TMD effective? How much would it cost?

Chinese policy seeks to further the development of China into the preeminent Western Pacific/Northeast Asian regional power.[21]Broadly stated, China believes missile defense programs will undermine strategic stability by weakening its one viable deterrent capability and precipitate a new arms race. Specifically, the Chinese worry that Japanese TMD programs will obviate their ballistic missile force; undermine their regional security; lead to Japan's eventual remilitarization; aid in the defense of Taiwan; and retard current global arms control efforts that reduce or check stronger powers such as the United States.

The United States would ideally like to participate in a bilateral theater missile defense that protects its forward deployed and forward based forces in Northeast Asia, that protects the Japanese main islands, and that does not precipitate a regional arms race. To this end, key U.S. concerns are proliferation of the ballistic missile threat, compliance with agreed-to nonproliferation protocols, and deployment of a viable TMD system without precipitating a

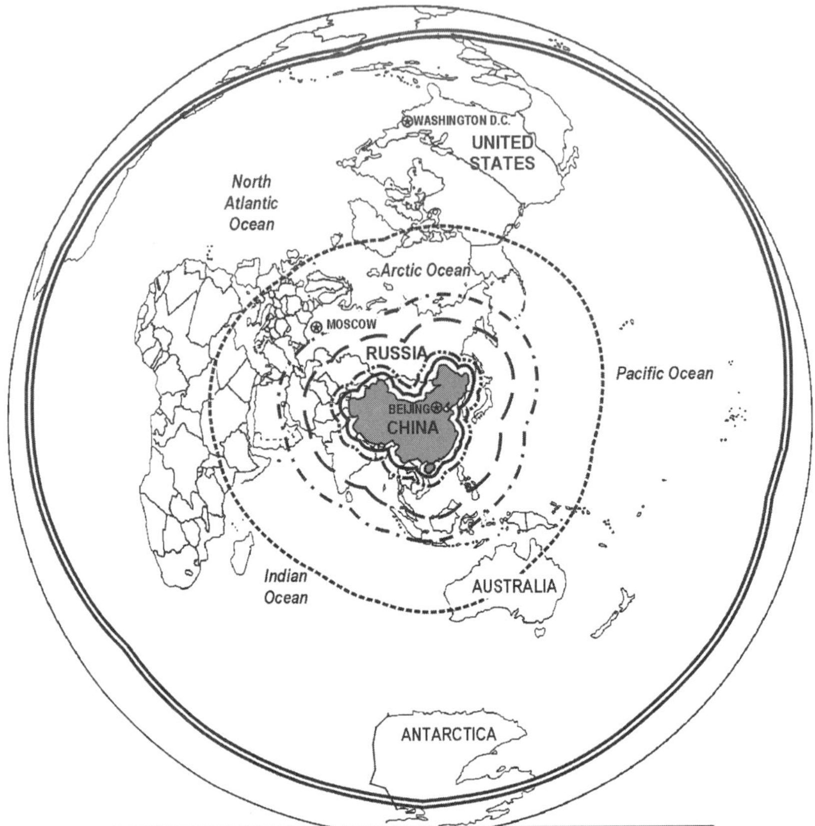

China

Current Missile Delivery System	Range (km)	Source
CSS-7	300	Domestic Production
CSS-6	600	Domestic Production
CSS-5	1,800	Domestic Production
CSS-2	2,800	Domestic Production
CSS-3	More than 5,500	Domestic Production
CSS-4	13,000	Domestic Production

Boundary representations are not necessarily authoritative

Figure 2. Estimated Ranges of Current Chinese Ballistic Missiles.[22]

cycle of improvements and counter-improvements in American and Chinese arsenals.

Considerations.

The United States and Japan estimate TMD will cost $15 billion for research, development and deployment based on the U.S. Navy Theater Wide system. But investment of this magnitude (approximately 15 percent of the total Japanese defense spending from 2000-2006) may adversely affect other necessary defense programs (such as FS-X, an indigenous fighter program). Fiscal constraint becomes a significant consideration in light of the stated Japanese policy of no increased defense spending over the next 5 years.

The U.S. side expresses concerns that the Japanese partnership may lead to redistribution of monies within the internal U.S. defense budget among competing missile defense programs. In other words, now that the Japanese have committed to cooperative research and development of an upper-tier naval system, the U.S. upper-tier land system may not receive the planned and budgeted amounts needed so that the partnership with Japan on the naval system can be kept on schedule.[23]

Japan is also very worried over Chinese reaction to its pursuit of TMD. Chinese officials encourage some of Japan's fears by maintaining that TMD in Japan would trigger a regional arms race. First, they argue TMD would require China to build up its own missiles to overcome the missile defense system; they warn that this "action-reaction" cycle could spin out of control.[24] Second, they claim Japan could use TMD technologies to manufacture offensive missiles. The Chinese would consider this very provocative, requiring a military response, most likely in the form of increased inventories of Chinese missiles.

For its part, the United States views the deployment of the North Korean Nodong-1 and Nodong-2 SRBM's and the

Missile System (U.S. Name in Brackets)	Range/Payload	Estimated Number in Service
DF-3 (CSS-2)	2,800 km/single nuclear or conventional warhead	38
DF-4 (CSS-3)	4,750 km/single nuclear warhead	12
DF-5/5A (CSS-4)	13,000+ km/single nuclear warhead	10
DF-21/21A (CSS-5)	1,800 km/single nuclear or conventional warhead	8
DF-15/M-9 (CSS-6)	600 km/single nuclear or conventional warhead	4
DF-11/M-11 (CSS-7)	300 km/single nuclear or conventional warhead	unknown
DF-31	8,000 km/single or MIRV nuclear warhead(s)	unknown
DF-41	12,000 km/single or MIRV nuclear warhead(s)	unknown

Table 4. Chinese Ballistic Missiles.[25]

development of the Taepo Dong-1 MRBM and Taepo Dong-2 ICBM (Table 3) as the most immediate and realistic threats in the region. Chinese CSS-5 and CSS-6 could also target Japan, but Taiwan dominates Chinese targeteering. Road mobile missiles under development in China (DF-31 and DF-41) (Table 4) are unlikely candidates for targeting Japan and by extension, U.S. forces, but they do retain the capability.[26]

The U.S.-Japan Alliance and Theater Missile Defense.

The current U.S.-Japan security relationship bases itself on the bilateral review that took place from October 1994 to April 1996, the resulting East Asia Strategy Report (Nye Initiative), and the follow-on U.S.-Japan Joint Security Declaration. This declaration attempted to demonstrate that the 2-year review of the security relationship had produced an alliance with great political and operational capability.[27]An initiative in the Joint Security Declaration called for a review of existing Guidelines for Defense Cooperation, seeking to expand them to include a focus on cooperation in responding to regional contingencies, as well as the direct defense of Japan. Concurrent negotiations quickly incorporated TMD cooperation into the Guidelines review process, which culminated in August 1999 when the United States and Japan signed a TMD cooperation agreement. This agreement calls for shared fiscal and technology responsibility in the development of a sea-based TMD.

U.S. Positions. U.S. proponents of TMD argue that, regardless of its impact on U.S.-China relations, or its effect on global arms control, or its effects on atomic states' nuclear strategy, TMD remains essential to defend deployed U.S. forces overseas. Proponents further claim that Japanese TMD offers the potential to enhance the bilateral security arrangement and provide funding, technology, and customers for an extremely costly and complex system.

Advocates maintain that TMD provides a bulwark for U.S. extended nuclear deterrence against newly emerging "rogue states," against emerging and established limited ballistic missile threats, and against chemical and biological threats for which nuclear deterrence alone does not offer sufficient deterrence. Second, TMD protects coalitions from intimidation. Third, TMD strengthens the reciprocal commitment for the United States to defend Japan and for Japan to defend forward-based U.S. forces. Fourth, assuming the U.S. deployment of TMD to protect its globally-based forces, TMD cannot be deployed to protect U.S. personnel without also affording protection to the Japanese population. Fifth, without TMD, Japan may feel more vulnerable to limited ballistic missile attack and may become predisposed to develop its own nuclear deterrent.[28]

Critics of Japanese TMD argue that the severity of the threat does not justify the cost. Further, they assert that TMD, as a tenant of counterproliferation, is faulty. TMD will overwhelm Japan's limited national security capabilities and leave Japan unprepared to handle the strategic responsibilities inherent to TMD. The cost of TMD is prohibitive for Japan. And, finally, critics assert that only arms reduction, not deterrence, guarantees security.

Japanese Positions. The advantages of TMD dominate the Japanese debate. The apparent lack of dissent results from the exceptional concern engendered by the North Korean Taepo Dong-1 missile flight test in August 1998 and the apparent benefit of TMD development to Japanese industry. Such solidarity will be temporary, however, as the high cost becomes clear.

Proponents of TMD stress that a viable missile threat exists from North Korea and China that must be contained. Second, they point out the potential for a significant amount of bilateral technology exchange with the United States, noting that industrial cooperation, to include licensed indigenous production of TMD components, will be worth billions of yen.[29]

Opponents of TMD argue that the prohibitive cost will not allow funding for other essential defense programs. They point out that upper-tier TMD systems prohibited by the 1972 ABM Treaty will limit the usefulness of TMD to Japan. Most importantly, adoption of TMD systems will precipitate confrontation with China.[30]

Alliance Implications. The complexity of TMD is staggering. Each partner must contemplate a broad range of detailed issues, not the least of which include formulation of an effective bilateral command and control strategy; the potential for truly integrated combined missile defense operations; and bilateral military connectivity that far outpaces what is now practical or even imagined in the alliance relationship.[31]

TMD will be effectively employed only if there exists a seamless and unimpeded command and control of disparate sensors and weapons, commanded by both nations, and controlled by a variety of interconnected military organizations, doctrinally linked; only if Japan's individual Self-Defense Forces cooperate with each other and with U.S. forces in unprecedented intimacy; and only if Japan's air defense system rebuilds from the ground up.[32]If and when all this happens, Japan will achieve a significant role in regional and global arms control and nonproliferation initiatives. Likewise, it will bolster the U.S. nuclear umbrella. What may start as an alliance management initiative may become an alliance-defining initiative, on the leading edge of the revolution in military affairs as part of a system that boasts unprecedented connectivity, interoperability and seamless command and control.[33]

Strategic Implications for the U.S.-China Relationship.

*TMD Transfers to Taiwan: "Umbrella over Taiwan."*The issue that arouses the most Chinese passion is Taiwan. Chinese nationalists vehemently claim Taiwan. They deflect any outside influence with charges of "interference in internal Chinese affairs" and "an affront to Chinese sovereignty." The rulers of the Chinese Communist Party (CCP) will address the Taiwan issue in terms of an "internal affair" and rebuff all attempts to legitimize other states' contact with Taipei. This approach serves, in turn, to uphold the legitimacy of the mainland regime and further isolate Taiwan from vital sources of economic, political, and military support.

China worries that if it wavers the slightest on Taiwan and the territory achieves viable independence, then other regions of China seeking autonomy, such as Tibet and perhaps Xinjiang, would be encouraged in their struggles. The specter of internal disruption always looms large in the CCP's leadership calculus.

The possibility of Japanese TMD equipment or technology being sold or transferred to Taiwan elicits a visceral, menacing reaction from the Chinese. China maintains that TMD technology and equipment transferred to Taiwan would exacerbate missile proliferation, violate the spirit and intent of the MTCR, and encourage Taiwanese ambitions for independence. Further, a sea-based Japanese TMD might also be used to protect Taiwan in the event of a military confrontation between Taiwan and mainland China, despite China's ability to overwhelm Taiwan with sheer numbers of missiles. The inclusion of Japan, and by extension the United States, in any cross-strait conflict dramatically raises the strategic stakes. This political issue, along with the potential for quick escalation to crisis proportions, causes the greatest concern for the Chinese, not actual TMD missile performance.

The importance of Taiwan in Chinese calculations about TMD cannot be overstated. The nature of the cross-strait conflict defies the usual arguments about the offense-defense balance and the "security dilemma." In sum, the arguments that the buildup of defensive weapons and adoption of defensive doctrine should not be destabilizing fall on deaf Chinese ears.

According to conventional wisdom, defensive weapons destabilize because they shore up the territorial status quo by deterring or physically preventing aggressors from achieving revisionist goals, whereas offensive weapons destabilize because they threaten the status quo.[34] China, however, retains as its primary policy objective the prevention of Taiwan declaring permanent independence from the Chinese nation. A cross-strait relationship that would legalize and freeze the territorial status quo remains the main threat to Chinese policy. China employs military and economic coercion as her primary means of countering that threat. Beijing considers traditionally defensive weapons in the hands of Taiwan and any potential political allies (the United States and Japan) as dangerous, because they give Taiwan officials additional confidence in their efforts to legitimize the territorial status quo.[35]

TMD in Japan would reduce China's ability to threaten Taiwan with ballistic missiles, her primary means of coercion. Significantly, the ship-based system under development causes acute worry because of its large "footprint" and easy deployability. Effectiveness of the system aside, China agonizes about the psychological and political effect the system would have on Taipei's attitudes about seeking diplomatic latitude, as well as U.S. and Japanese attitudes about cross-strait relations.[36]

TMD Cooperation Will Lead to Japan's Eventual Remilitarization. The Chinese perceive a shift in the U.S.-Japan alliance dating back to the mid-1980s. They maintain that *de facto* licensed coproduction programs such as the F-15J and Patriot PAC 2+, coupled with changes in alliance structure precipitated by the Nye Initiative and the subsequently revised Defense Guidelines, show a willingness, perhaps even an eagerness, by the United States and Japan to "normalize" Japan's security.

TMD heightens this concern when viewed in the aggregate with other U.S.-Japan alliance developments. These include the new Japanese LST (landing ship, tank), a ship capable of launching and recovering aircraft; the family of Japanese space launch vehicles (SLV) regarded as potential dual-use platforms; and the Japanese purchase of Patriot PAC-3 missiles with their limited lower-tier capability.[37] China views such developments as precursors to the relaxing of Japanese law.[38]

Indeed, U.S. pressures on Japan to accept a greater share of the burden of its own national defense in effect encourages Japan to become self-sufficient. Further, in the 56 years since the end of World War II, it has become inevitable that the post-war U.S.-Japan security relationship will change, especially in view of Japan's growth as a global economic power. Wealth begets the imminent potential for military power.

TMD and the prospect of meshing national command and control systems, both for regional defense and as an "early warning" for proposed U.S. NMD, mean that Japan will develop capabilities beyond what its present constitution allows.[39] Complementary to TMD is a space-based sensor system to aid in early warning and detection of incoming missiles. Japanese law prohibits the militarization of outer space. Satellite sensors to warn of a hostile missile launch and perhaps to help guide an intercept constitute a necessary part of a TMD system.

China recognized these overt harbingers of change. It also recognized the implications of the Japanese Ministry of Foreign Affairs future diplomatic blueprint, Challenge 2001.[40] Accordingly, the Chinese assert that TMD represents another, albeit significant, step by Japan to take its national security into its own hands. The implications for China of a militarized, self-sufficient Japan defy description. Japan's technical ability, industrial potential, and geographic position would inhibit the Chinese desire to remain militarily preeminent in Northeast Asia. The most discussed development of any "remilitarization" of Japan and the TMD implications of that action concern an independent Japanese ability to build and deliver offensive ballistic missiles. China maintains that the technology transfers of TMD components are directly applicable to offensive systems. Hence, China views U.S. cooperation with Japan on TMD as nothing less than U.S. support for emergent Japanese military power. In response, the United States reassures that all MTCR regimes would be followed. The United States contends that, in fact, Japan's indigenous space program and its family of SLVs exhibit more compatibility for this type of technology transfer than TMD.[41]

TMD Will Lead to an Independent Japanese Nuclear Deterrent. The Japanese debate on security normalization and nuclear weapons development is in its infancy and the implications of self-sufficiency remain unspoken. As the only nation to endure nuclear attack, Japan can articulate the most compelling reasons for not developing such a capability.

But China maintains that Japan revealed its pragmatism regarding the use of nuclear devices when it proceeded with the commercial development of nuclear power in the 1970s. Perceptual differences here again take on exceptional significance. Old hostilities from the war years of the 1930s and 1940s do not allow for a Chinese acceptance of Japan's benign use of nuclear power absent U.S. restraints.

More pragmatically, however, Japanese nuclear power exhibits remarkable transparency (transparency addresses the ability to verify that materials are not used for nuclear weapons making, etc.). So mutual confidence measures and transparency exercises could keep the "nuclear genie in the bottle" and limit regional fears. Further, TMD obviates the need for an independent Japanese deterrent by shoring up the U.S. nuclear deterrent.[42]

TMD Undermines China's Nuclear Deterrent. This is the central issue to the U.S.–China missile defense debate. The United States relies on both offensive forces (the strategic nuclear triad of land-based missiles, sea-based missiles, and bombers) and defensive forces (ballistic missile defenses), with the latter providing reinforcement to the stabilizing influence of deterrence. China derives its deterrence solely from an offensive land based missile force.

These differences are stark. China's reliance on only one form of nuclear deterrence makes any attempt to restrain it destabilizing. China consistently maintains TMD undermines its offensive nuclear deterrence. China's declared no-first use (NFU) of nuclear weapons contradicts this argument, however, unless its declared policy is false and China, in fact, targets Japan (active targeting is considered a violation of NFU protocols). So China's objections to TMD reveal a significant inconsistency in Chinese doctrine.[43]

TMD Will Provoke Changes in Chinese Nuclear Doctrine. Chinese responses to Japanese TMD suggest support for the evolution of Chinese nuclear doctrine from one of "minimum deterrence" to one of "limited deterrence." Minimum deterrence, according to China, means possession of a small number of nuclear warheads sufficient to inflict unacceptable damage on a handful of enemy cities.[44]In truth, Chinese missiles can presently do little else. As the potential of Chinese weapons systems increases, however, there exist indications that the Chinese are building a more robust limited nuclear deterrent. Limited nuclear deterrence (in concert with the technological and economic capability of the Chinese state) originally envisioned the ability to inflict unacceptable damage with several warheads aimed at enemy cities. It since matured into a more sharply defined concept that now promises limited counterforce warfighting capabilities. Limited deterrence now means having enough capability to deter conventional theater and strategic nuclear war and to control and suppress escalation during a nuclear conflict.[45]

This shift invites several questions. Has the Chinese nuclear stockpile expanded beyond the commonly accepted 75-100 warheads available for targeting against Taiwan (and Japan) to a force that is more consistent with a limited deterrence capability (250-300 warheads)? We do not know, but remain convinced it has. The exceptional circumspection of the Chinese about the size and capability of their ballistic missile force breeds mistrust. But doesn't this shift to greater capability justify TMD? Japan remains unconvinced of benign Chinese intentions, while China insists its declared NFU policy and strictly defensive nuclear weapons offer indications of peaceful intent that should be accepted at face value.

China's threat of building to "limited" nuclear deterrence is a thinly veiled suggestion that Japanese TMD may provoke an even greater increase in the number of Chinese missiles, the method by which they are deployed, and the technical capabilities of the missiles (addition of Multiple Independent Re-entry Vehicles [MIRV], decoys, penetration aids, etc.). China developed and deployed its first generation of missiles in 11 years.[46]The second generation will take somewhat longer. Forecasted with an initial operating capability during the year 2000, China's move to modernize occurred well before TMD became a significant issue.

On the other hand, it can be argued that China is progressing towards nuclear modernization anyway. It came to the conclusion to do so independent of, or minimally affected by, TMD considerations. Thus all the rhetoric about U.S.-Japan TMD could be viewed as a ploy to justify the expansion of China's nuclear arsenal, which would happen anyway.

TMD Undermines Arms Control. The most articulate expression of Chinese sentiments comes from Ambassador Shu Zukang, who asserts that "the U.S. development of missile defenses does not contribute to global stability nor will it serve the interests of the U.S. itself."[47]He contends that NMD/TMD violates the 1972 ABM treaty, considered by the Chinese as "the most important arms control protocol being implemented," and that the loss of this treaty will lead to a disruption of the strategic balance between the United States and Russia.[48] This will, in turn, result in a greater instability for all countries. Specifically, Sha asks, "How can you expect progress in arms control and nonproliferation while you yourself are developing NMD at full steam? It is just wishful thinking."[49] Shu also hinted that rather than engaging in an arms race, should the United States proceed with its NMD/TMD deployments, "it would be quite possible for China to review its policies on various arms control, disarmament and nonproliferation issues, including FMCT negotiations."[50] Chinese officials also assert that TMD deployments will lead to a complete halt in global arms control and nonproliferation, with direct implications for START II, the Comprehensive Test Ban Treaty, Missile Technology Control Regime and Fissile Material Cut-off Treaty. Chinese officials have gone so far as to declare "NMD will also cause many nations to build up their nuclear forces and further retard the disarmament process."[51]

Significantly, China avoids any overtures to participate in multilateral ABM controls and continues to upgrade Soviet designed surface-to-air (SAM) missiles for the ballistic missile defense role. Since the U.S.-Russian agreement on threshold testing of missile defense systems clarified permitted capabilities, Chinese arguments about undermining the ABM Treaty have lost their relevance.[52]

China shows deep concern that should TMD/NMD undermine the ABM Treaty, then Russian ratification of START II and participation in START III would be obviated, with negative consequences for Beijing. Specifically, by withdrawing from the START III preliminaries, the Russian threat to the northern Chinese border would return, albeit on a smaller scale than during the 1960s to 1990s rift between the two. This would divert Chinese resources from development and modernization and inhibit the current free flow of technology and capital between them.[53]

TMD missile legitimacy also received a challenge under the 1987 MTCR. Some states construed TMD development as a form of missile proliferation. The United States maintains that TMD development protocols do not constitute a violation of the MTCR and will not contribute to missile proliferation.[54]

For its part, the United States maintains that it is not rushing toward upper-tier TMD deployment in Japan and that it remains undecided about selling lower-tier TMD to Taiwan. The only TMD deployment decision the United States has made addresses lower-tier TMD deployments in Japan to protect U.S. troops and facilities. Taiwan cannot "plug into" an existing TMD system, thus Chinese views of TMD should not be based on the misperception that the United States seeks to build a regional TMD architecture.[55]

Possibilities.

The United States has four viable options in regard to Northeast Asia theater missile defense. First, the United States and Japan should continue to follow a path to eventual deployment of a viable TMD system in Japan and South Korea. The essential U.S. obligation to protect its forward-based military personnel demands an effective TMD in these two countries. Anything less could be construed as gross negligence of a country's obligation to adequately protect its forces from a known, viable threat that is economically and politically feasible to address. Likewise, intermingling of the American and Japanese population in Japan renders it impossible to protect one without extending that protection to the other.

The alliance partners, however, should be exceptionally cognizant of the remaining alternatives that would make this bilateral choice more palatable to China and thus avoid escalating the arms race. The second alternative would be to pursue THAAD, rather than NTW. If effective, this could provide the home islands and the U.S. forces protection. Yet, THAAD's relative immobility and inability to extend its "footprint" to Taiwan would be less likely to exacerbate Chinese concerns about real or perceived alliance support for Taiwanese independence.[56] Of course, this would raise new U.S.-Japan concerns, particularly with cooperative development and less Japanese industrial participation in a more mature system. But it may be worth this price not to overtly antagonize Beijing.

A third choice, less likely since August 1999, is production of TMD unilaterally by the United States. Chinese analysts have consistently pointed out that U.S.-Japan coproduction of TMD carries a fundamentally different and more provocative political meaning for China than if the United States produced such systems without Japanese help as part of the U.S. strategy of protecting its forward deployed forces. This directly challenges the "protect us, protect them" pillar of U.S.-Japan TMD partnering.

The fourth option calls for no deployment of TMD at all. Proponents stress that this would obviate the Chinese arguments about arms control and nonproliferation and put real pressure on Beijing to be genuinely transparent about missile development and deployment. Reconsideration of deployment might also open up an opportunity to exploit Chinese concerns and encourage Beijing to participate in a formal trilateral security dialogue.

In contrast, the very real and more likely possibility envisions continued Chinese obfuscation on its military affairs. The overwhelming desire to build, deploy, and maintain the military trappings of a "great power" remains so engrained in the Chinese leadership's psyche that any risky attempt at placating Chinese fears with genuine measures of conciliation would more than likely be fruitless. Given the Taiwan problem and overwhelming superiority of U.S. military power, China shows the deep-seated desire to develop its missile capability to a significant degree regardless of U.S.-Japan TMD cooperation.

Conclusions.

By persistently increasing U.S. capabilities and judiciously incorporating increased Japanese capabilities in the U.S.-Japan alliance, the United States can fulfill its obligation to its forward-deployed forces and allies in Northeast Asia. The United States needs to be cautious, however, about what new roles Japan plays in the alliance, for TMD will fundamentally change that relationship. The ultimate responsibility of an American President is to provide for the safety of the American people and the armed forces by deterring attack on the homeland and allies. By increasing U.S. capabilities, the United States will be better able to handle future regional crises and prevent them from happening. Consistent actions in the face of inflammatory rhetoric will avoid the destabilizing effects of an arms race and backsliding on arms control and nonproliferation agreements. Ultimately, such actions will contribute mightily toward long-term peace and stability in Northeast China.

ENDNOTES

Editor's Note: For the reader/researcher seeking additional information on this subject, please review the following published article: Thomas J. Christensen, "Theater Missile Defense and Taiwan's Security," *Orbis*, Winter 2000, pp. 79-90.

1. Barbara Opall-Rome, "One on One," *Defense News*, February 1, 1999, p. 22.

2. "Missiles, Theater Missile Defense, and Regional Stability. Second U.S.-China Conference on Arms Control, Disarmament and Nonproliferation," linked from the Center for Nonproliferation Studies at the Monterey Institute of International Studies; available from http://www.cns.miis.edu/cns/projects/eanp/conf/uschina2/index.htm; Internet; accessed December 7, 1999, p. 2.

3. *Ibid.*, p. 3.

4. "Japan Moves," *China Daily*, November 18, 1998, p. 4.

5. "Pyongyang's Missile Program Threatens Area Peace," *Indian Express*, October 9, 1998, sec. A, p. 1.

6. Patrick Cronin, Paul Giarra, and Michael Green, "Alliance Implications of Theater Missile Defense," in *The U.S.-Japan Alliance: Past, Present, and Future*, Patrick Cronin and Michael Green, eds., New York: Council on Foreign Relations, 1999, pp. 171-172.

7.Cronin, Giarra, and Green, p. 172.

8. "Fact Sheets for Ballistic Missile Defense Systems" linked from the Ballistic Missile Defense Organization available from http://www.acq.osd.mil/bmdo/bmdolink/meads.html; Internet; accessed November 30, 1999.

9. Cronin, Giarra, and Green, p. 172.

10. Robert Wall, "U.S.-Japan Agree on Cooperative Missile Defense," *Aviation Week and Space Technology*, August 23, 1999, p. 48.

11. Cronin, Giarra, and Green, p. 172.

12. *Ibid.*, p. 173.

13. *Ibid.*, p. 50.

14. Wall, p. 48.

15. Office of the Secretary of Defense, *Proliferation: Threat and Response*, Washington, DC: U.S. Department of Defense, November 1997, p. 7.

16. Thomas J. Christensen, "China, the U.S.-Japan Alliance and the Security Dilemma in East China," *International Security*, Vol. 23, Spring 1999, p. 49.

17. *Ibid.*

18. *Ibid.*, pp. 49-50.

19. *Ibid.*, p. 50.

20.Paul Giarra, "Interim Report from an Evolving Alliance—Japan's Changing Security Policy: Listening All These Years," chapter from unpublished book, furnished December 1999.

21. "Missiles, Theater Missile Defense . . .," p. 13.

22. Office of the Secretary of Defense, *Proliferation: Threat and Response*, p. 11.

23. The ideas is this paragraph are based on remarks made by a speaker participating in the Commandant's Lecture Series at the Army War College, AY 1999-2000.

24. Giarra, p. 10.

25. "Chinese Ballistic Missiles" linked from the Center for Nonproliferation Studies at the Monterey Institute of International Studies; available from *http://www.cns.miis.edu/cns/projects/eanp/pubs/chinanuc/bmsl.htm*; Internet; accessed December 5, 1999.

26. Office of the Secretary of Defense, *Proliferation: Threat and Response*, p. 10.

27. Cronin, Giarra, and Green, p. ix.

28. *Ibid*., p. 174.

29. *Ibid*., p. 177.

30. *Ibid*.

31. *Ibid*., p. 180.

32. *Ibid*., pp. 181-182.

33. *Ibid*.

34. Christensen, pp. 64-65.

35. *Ibid*., p. 66.

36. *Ibid*.

37. *Ibid*.

38. *Ibid*.

39. Michael J. Green, "Theater Missile Defense and Strategic Relations with China," in *Restructuring the U.S-Japan Alliance: Toward a More Equal Partnership*, Ralph A. Costa, ed., Washington, DC: CSIS Press, 1997, p. 116.

40. Green, p. 116.

41. Green, pp. 116-117.

42. *Ibid.*, p. 117.

43. *Ibid.*, p. 115.

44. Alastair I. Johnston, "China's New 'Old Thinking,'" *International Security*, Vol. 20, Winter 1995/1996, p. 10.

45. *Ibid.*, p. 19.

46. *Ibid.*, p. 36.

47. "Missiles, Theater Missile Defense . . .," p. 4.

48. *Ibid.*, p. 14.

49. *Ibid.*, p. 4-5.

50. *Ibid.*, p. 5.

51. *Ibid.*

52. Green, p. 116.

53. "Missiles, Theater Missile Defense . . .," p. 14.

54. Evan A. Feigenbaum, "Who's Behind China's High-Technology 'Revolution'?," *International Security*, Vol. 24, Summer 1999, p. 104.

55. *Ibid.*, p. 13.

56. Christensen, p. 76.

GLOSSARY

CTBT	Comprehensive Test Ban Treaty. International treaty whereby the signatories agree not to conduct nuclear device testing of any type.
EPCI	Enhanced Proliferation Control Initiative. An initiative that enables the U.S. Government to require export license of all items, even those not on the control list in the Export Administration Regulations, if the exporter knows, has reason to know, or has been informed that the items will be used in a nuclear missile.
Endo-Atmospheric	Within the earth's atmosphere. Generally regarded as less than 60 miles altitude.
Exo-Atmospheric	Outside the earth's atmosphere. Generally regarded as greater than 60 miles altitude.
FMCT	Fissile Material Cut-off Treaty.
JTMD	Joint Theater Missile Defense. The integration of joint force capabilities to destroy enemy theater missiles in flight or prior to otherwise disrupt the enemy's theater missile operations though an appropriate mix of mutually supportive passive missile defense; active missile defense; attack operations; and supporting command, control, communications, computers and intelligence measures.

Lower Tier	Those missile defense systems developed for engaging theater ballistic missiles in an endo-atmospheric role, usually in the descent phase of flight. Missiles engaged are usually shorter range and lower speed (warheads traveling < 5 km/sec)
MTCR	Missile Technology Control Regime. An informal, non-treaty association of states that have an established policy or interest in limiting the spread of missiles or missile technology and of limiting the risks of proliferation of weapons of mass destruction by controlling transfers that could make a contribution to the delivery of such weapons.
NAW	Navy Area Wide. A U.S. Navy lower-tier missile defense system under development by the United States. Deployed from Aegis-class cruisers, it uses a modification of an existing surface-to-air missile to engage theater ballistic missiles.
NMD	National Missile Defense. A U.S. Program, run by the Department of Defense Ballistic Missile Defense Office (BMDO) tasked with developing a system(s) that will protect the continental United States from ballistic missile attack.
NPT	Non-Proliferation Treaty. A treaty designed to limit the formation of multinational nuclear forces and prevent the uncontrolled spread of nuclear weapons.

NTW	Navy Theater Wide. A U.S. Navy upper-tier missile defense system under development by the United States and Japan. Deployed from Aegis-class cruisers, it uses a modification of an existing surface-to-air missile to engage theater ballistic missiles
PAC-3	Patriot Advanced Capability 3. An upper-tier missile system under development by modifying a version of the Patriot surface-to-air missile for intercepting theater ballistic missiles.
SLV	Space Launch Vehicle. Those vehicles whose primary purpose is to launch into space with cargoes other than weapons.
TBM	Theater Ballistic Missile. A missile whose target may be within a given theater of operations.
THAAD	Theater High Altitude Air Defense. A U.S. Army upper-tier missile defense system that utilizes an exo-atmospheric "hit-to-kiss" missile and warhead to intercept theater ballistic missiles.
Upper-tier	Those missile defense systems developed for engaging missiles in an exo-atmospheric role, either in the boost, ascent, or descent phase of flight. Missiles engaged are usually longer range and higher speed (warheads traveling > 5 km/sec)